ONE GIANT LEAP FOR MANKIND

REVISED EDITION

Turning Points in American History

ONE GIANT LEAP FOR MANKIND

Carter Smith III

Silver Burdett Company, Morristown, New Jersey

Cincinnati; Glenview, Ill.; San Carlos, Calif.;
Dallas; Atlanta; Agincourt, Ontario

Acknowledgements

We would like to thank the following people for reviewing the manuscript and for their guidance and helpful suggestions: Professor Kenneth Kusmer, Department of History, Temple University; and Diane Sielski, Library Media Coordinator, Coldwater Village Exempted Schools, Coldwater, Ohio. We would also like to thank Marie Jones of the Photo Archives division of the National Aeronautics and Space Administration for her help and cooperation.

Cover: Edwin R. ("Buzz") Aldrin plants the flag of the United States on the surface of the moon
Title page: Earth rising over the moon
Contents page: Gemini 7 soaring over the earth

All photographs courtesy of the National Aeronautics and Space Administration unless otherwise indicated.

Library of Congress Cataloging-in-Publication Data

Smith, Carter, 1962-
    One giant leap for mankind.

    (Turning points in American history)
    Summary: Discusses the developing events taking
rockets, satellites, and man into space.
    1. Outer space — Exploration — Juvenile literature.
[1. Outer space — Exploration] I. Title. II. Series.
TL547.S38 1989   919.9′04   89-32378
ISBN 0-382-09909-5
    0-382-09910-9 (pbk.)

 Created by Media Projects Incorporated

Series design by Bruce Glassman
Bernard Schleifer, Design Consultant
Ellen Coffey, Project Editor
Jeffrey Woldt, Photo Research Editor
Charlotte A. Freeman, Photo Research Associate

Published simultaneously in Canada by GLC/Silver Burdett Publishers

Manufactured in the United States of America

CONTENTS

INTRODUCTION

THE EAGLE HAS LANDED

In the early morning of July 16, 1969, the astronauts of Apollo 11, Neil Armstrong, Edwin "Buzz" Aldrin, and Mike Collins, sat down for a breakfast of steak and eggs, toast, juice, and coffee. This was no ordinary day. For these three men—indeed, for an awaiting world—it was the day that a tiny capsule on top of a powerful rocket would take the three Americans off into space and on to a place people had dreamed of going to ever since they had first looked up and seen the stars. Their destination: the moon.

After breakfast Armstrong, Aldrin, and Collins went up to the suit room to begin the long process of being fitted into their space suits, a job so complicated that a fourth man had to be there to supervise and assist them. Finally, suited up, the three made their way through the throngs

Apollo 11 on the launch pad

of reporters and cameramen and climbed into a waiting van that took them to the launch pad. Already on the pad was the giant Saturn V rocket that would hurtle them into space. Gripping the rocket, which was taller than a football field set on its end, was a huge steel scaffold. Known as the "launch umbilical tower," this scaffold was designed to hold the rocket until the last second. At the tower's base was an elevator, its doors open, ready to whisk the astronauts to the top of the rocket where the command module, Columbia, and the lunar module, Eagle, were located. When the elevator reached the hatch of the Columbia, 320 feet above the ground, the three astronauts took a last look at the nearly twenty thousand VIPs and members of the press and other spectators who had gathered below. Then Armstrong stepped through the open hatch, followed by Collins and Aldrin. After some final systems checks, a member of the backup

The crew of Apollo 11 photographed in front of the test module

team carefully strapped them all in, then left, closing the hatch behind him. It would be eight days before the three astronauts would set foot again on Earth.

At last the countdown began. At precisely 9:32 A.M. Eastern time, the crew of Apollo 11 blasted off. According to Buzz Aldrin, they didn't even know they were moving until they heard the voice of Launch Control announce it. But the spectators below knew, and they began cheering and shouting.

After four busy days in flight, which included separating the command module from the rest of the rocket and docking it with the Eagle, the astronauts were ready for the final phase of their mission—the actual descent to the moon. Even though Collins, who was the command-module pilot, would remain on board to steer the Columbia as it orbited the moon, he was as busy as the other two in preparations for the lunar-module landing.

Finally, they were ready. Aldrin and Armstrong entered the Eagle. First they sealed the hatch that led from the Columbia. Then they deployed the Eagle's spider-like landing gear. After getting confirmation from Mission Control in Houston, they undocked the Eagle from its command module.

Still about sixty miles from the moon's surface, the Eagle began drifting downward, coasting toward the near side of the moon. At exactly the right moment, about twelve minutes from the actual landing, the Eagle's engine came on. But then, all of a sudden, when they were six thousand feet from the surface, a yellow warning light flashed on, signaling that the Eagle's computer had overloaded. Luckily, the flight controller responsible for the lunar module's computer system, Steve Bales, had been monitoring the descent closely from Houston. In a split-second analysis of the problem, he decided that it had happened because the people who had designed the computer had never met with those who had designed the landing radar. Therefore, since the computer failed to receive proper signals from the landing gear, the system overloaded. After reassuring the two Eagle astronauts, Bales told them to proceed with the descent.

As the descent continued, Armstrong

Edwin R. ("Buzz") Aldrin steps onto the lunar surface

took over the controls at five hundred feet to prevent the Eagle from landing in the middle of a dangerous, rock-strewn crater. Meanwhile, with people around the world listening on their radios and TVs, Aldrin described the descent to Mission Control: "Seventy-five feet, things looking good . . . lights on . . .kicking up some dust. Thirty feet . . . drifting to the right a little . . . contact light . . . okay. Engine stop."

The next voice was Commander Armstrong's: "Houston, Tranquillity Base here. The Eagle has landed," he said calmly.

In fact, the lunar module had landed at exactly the spot the mission had called for. Shortly before 11 P.M., the astronauts opened the hatch. As mission commander, Armstrong was the first out. On the second rung of the Eagle's ladder, he pulled a ring at the side of the spacecraft, deploying a television camera. Then, as millions of people back on Earth watched breathlessly, he planted his left foot on the surface of the moon. And as he did, he said, "That's one small step for man . . . ah . . . one giant leap for mankind."

1

"THE WILL TO DO IT"

Perhaps the greatest fantasy that humans have ever had was that someday they would be able to fly. These days jet airplanes and helicopters are a part of everyday life, so it is easy to forget that they have not really been around long at all. But long before the first airplane ever flew, people were dreaming about flying and—even more daring—about leaving the earth's atmosphere altogether. The ancient Greeks, for example, told a story about a foolish young man named Icarus who fell from the sky after the wings he had made out of wax melted when he flew too close to the sun.

We have come a long way since Icarus. But that would not be so if stories like his had not challenged us to experiment with flight and explore the skies. One of the

Diagram of the universe with Earth as the center

earliest astronomers was a second-century Greco-Egyptian named Ptolemy. He believed that the earth was at the center of the universe, and for hundreds of years this idea was accepted as truth. It was not until the fifteenth century that Nicolaus Copernicus stated his theory that the earth actually orbited around the sun. By the time the Italian astronomer Galileo Galilei developed the telescope in the early seventeenth century, the concept of an Earth-centered universe had been abandoned.

With the telescope, humans could begin to get an accurate picture of just what our solar system looked like. Until 1781, however, we had discovered only six planets. Then, while his country was busy fighting a losing battle against her colonies in America, Englishman William Herschel discovered a seventh, Uranus. The eighth planet, Neptune, was discovered in 1846, but it wasn't until 1930 that a young lab assistant named Clyde Tombaugh, working in The Lowell Laboratory in Flagstaff, Arizona, found the ninth and so far the farthermost planet from the sun, Pluto.

While the astronomers studied the stars, many writers created fantastic stories about how to get to them. Back in the seventeenth century, Cyrano de Bergerac, perhaps the earliest science-fiction writer, wrote about a machine powered by rockets that blasted him up "into the clouds," introducing the idea of the use of rocketry for space exploration.

During the 1860s, the French writer Jules Verne described steering rockets used to maneuver his spaceship in *From the Earth to the Moon*. Verne's ship took off

An illustration of the rocket "Columbiad" from Jules Verne's From the Earth to the Moon

from a 900-foot-long cannon buried in the sands of southern Florida. The first NASA launches actually took place in southern Florida about a hundred years later, and Verne's idea of steering rockets is one that influenced early rocket engineers.

Many historians believe that rockets were first invented by the Chinese nearly a thousand years ago, soon after they had discovered gun powder. In A.D. 1232, when the Mongols attacked the Chinese city of Kai-feng Fu, the Chinese defenders used weapons described as "arrows of flying fire." During this battle, the Chinese also dropped from the walls of their city bombs described as "heaven-shaking thunder." It is believed that both of these weapons were actually primitive rockets made from gunpowder. The Mongols eventually carried this new rocket weaponry to the Near East and on to Europe. By the early sixteenth century, many Europeans had become fascinated by them. Indeed, as early as 1500, the great artist Leonardo da Vinci sketched his conceptions of rocketlike missiles.

Because of their potential for use in warfare, rockets were developed very quickly in the nineteenth century. The British launched more than 25,000 rockets against the Danish city of Copenhagen in 1807. Just a few years later, the same kind of rockets, each weighing about thirty pounds, were used against the United States in the War of 1812. The words "the rockets' red glare" in Francis Scott Key's "Star-Spangled Banner" refer to British rockets launched in the attack on Baltimore's Fort McHenry.

During the First World War, the military use of rockets continued. They were employed mainly to launch the flares that revealed enemy positions at night. But with the advent of the airplane, rockets were needed for this purpose less and less.

Still, during the twentieth century, the science of rocketry grew as never before. Most of the important research and discoveries must be credited to a handful of pioneers. The most important of these were Konstantin Tsiolkovsky, a Russian; Hermann Oberth and Wernher von Braun, both Germans; and Robert H. Goddard, an American.

Konstantin Tsiolkovsky, born in 1857, suffered from a loss of hearing owing to childhood scarlet fever. Unable to hear the outside world, he turned instead to dreaming and fantasy. He wrote in his diary, "In my imagination, I could jump higher than anybody else, climbed poles like a cat, and walked ropes. I dreamed . . . there was no such thing as gravity."

Such dreams led Tsiolkovsky to experiment with the laws of gravity. Soon he was building balloons, a "mechanical hawk," and an all-metal dirigible. Not surprisingly, young Tsiolkovsky admitted that he was greatly influenced by Jules Verne's science-fiction stories. He wrote, "To place one's feet on the soil of asteroids, to lift a stone from the moon with your hand, to observe Mars at a distance of several tens of miles, to descend to its satellites and even to its own surface—what could be more insane! However, only at such a time when reactive devices [rock-

Konstantin Tsiolkovsky

ets] are applied will a new great era begin."

In 1903, Tsiolkovsky published an important article on rocketry in the journal *Scientific Review*. Many of his ideas broke new ground, setting the stage for later experiments. Some of those ideas, such as his multi-stage rocket and his methods of cooling the rocket's combustion chamber, are crucial to rocketry even today. By the time he died, in 1935, this pioneer had written thousands of scientific papers as well as a number of science-fiction books.

Though the advances in rocketry were made by just a few pioneers, fascination with rockets was shared by millions. Dur-

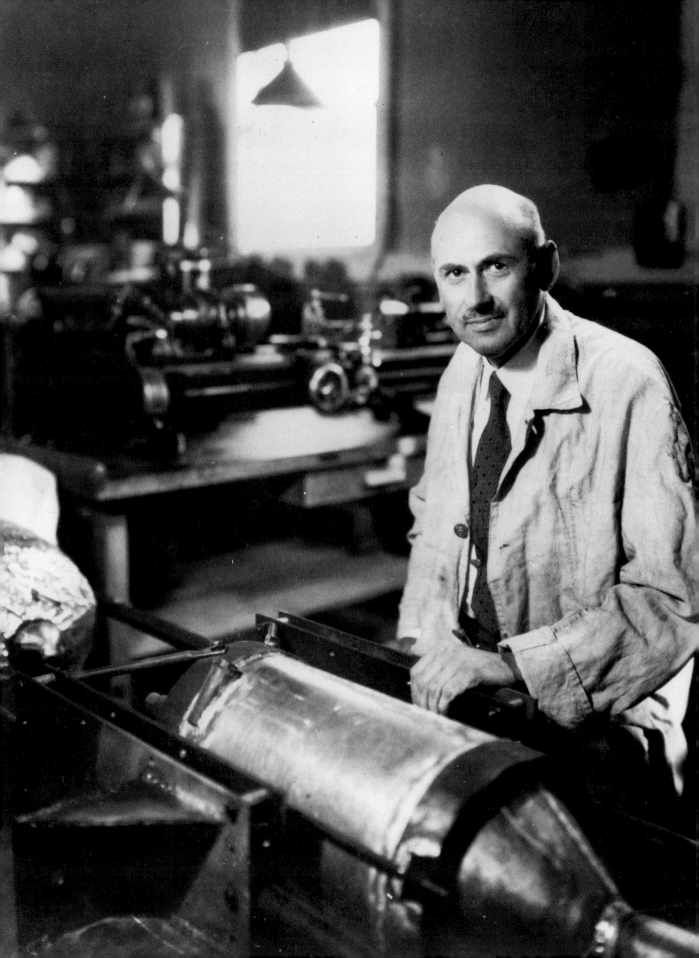

ing the late 1920s, many hobbyists in the United States, the Soviet Union, England, and Germany had formed amateur rocketry societies. Even as the Great Depression arrived, people began to escape earthly problems by turning to science fiction. Comic-book heroes like Flash Gordon and Buck Rogers took fans along as they traveled through Earth's galaxy and beyond. And just as Tsiolkovsky learned from Jules Verne, more recent rocketeers have learned from these space heroes. In fact, the helmets that the Gemini astronauts of the mid-1960s wore were developed from the design of a helmet first worn by Flash Gordon.

The American Robert H. Goddard began to write papers on space while he was still in high school, and when he graduated from Clark University, in Worcester, Massachusetts, in 1911, he already knew that one of the best ways to make rocket propellant was by combining liquid oxygen and liquid hydrogen. Soon afterward, he conducted successful tests of various rockets, including several powder-propelled models that rose to heights of nearly 500 feet.

In 1917, Goddard received his first big break. The Smithsonian Institution gave him $5,000 to do a study to determine just how high a rocket could go. After two years he released his report. Called "A Method of Reaching Extreme Altitudes," it indicated that, in theory at least, humans could send a rocket to the moon.

Robert H. Goddard

Goddard suggested that in order to tell if the rocket had reached its target, flash powder could be added to the missile so that it could be seen with a telescope when it exploded on impact. It was this latter idea that immediately attracted attention, not the more important rocket theory. Nonetheless, Goddard suddenly found himself famous.

More interested in his work than in his fame, however, Goddard quietly continued his research. In March 1926, a twelve-foot rocket, the first ever to use a liquid propellant, took off from his Aunt Effie's farm, near Auburn, Massachusetts. It was a very short flight, rising to a height of 41 feet and traveling a distance of 184 feet at a speed of 60 miles per hour. Still, it was the first flight of its kind and one of Goddard's most significant advances.

All this time, the newspapers had been following Goddard's work, and another famous man in the history of flight grew more and more interested in what Goddard was doing. Finally, in November 1929, Charles A. Lindbergh, the first person to fly across the Atlantic Ocean alone in a plane, paid Goddard a visit. Lindbergh was so impressed that he persuaded philanthropist Daniel Guggenheim to give Goddard $100,000 to set up a facility in New Mexico where he could continue his studies.

Soon afterward, near Roswell, New Mexico, Goddard launched a specially designed 85-pound missile that rose 7,500 feet into the air. A second rocket reached a speed that nearly broke the sound barrier.

While Goddard continued his work, with more help from the Guggenheim Foundation, his name and reputation crossed the Atlantic to Germany. There, Hermann Oberth had been experimenting with liquid-propelled missiles, and he wrote to Goddard, suggesting that they keep in touch. In one letter he said "I think that only by the common work of scholars of all nations can this great problem be solved . . . to pass over the atmosphere of our earth by means of a rocket."

In 1923, Oberth wrote a book called *The Rocket into Planetary Space*, in which he proposed the design for a liquid-propellant rocket capable of exploring the outer reaches of the atmosphere. Six years later, he wrote a follow-up entitled *The Road to Space Travel*. That year he was named president of Germany's Society for Space Travel.

By 1932, the Society for Space Travel had established a testing ground at Kummersdorf, outside Berlin. One of Oberth's students, Wernher von Braun, persuaded the German army to invest a small amount of money in rocketry. Two years later, von Braun and his colleagues launched two liquid-fueled projectiles to a height of 6,500 feet. By the time they had moved to a new site, at Peenemünde, on the coast of the Baltic Sea, they had carried out close to one hundred such tests.

While German scientists conducted experiments at Kummersdorf and Peenemünde, the scientists of the Soviet Union had been conducting tests of their own. During the early 1930s, they had launched rockets up to 3.5 miles into the sky. And by 1940, they had found that a combination of liquid and solid propellants would allow rockets to travel as high as 12 miles.

With the arrival of World War II, however, progress in rocketry took a back seat to the demands of war. Goddard went to work for the U.S. Army. In Germany, von Braun ran into problems getting financial backing for his experiments. Eventually, the German rocket program was given money by the government, and German scientists began to develop such weapons as surface-to-air missiles, air-to-surface missiles, and air-to-air missiles.

But the most important developments, as well as the most deadly in war, came from Peenemünde. In 1943, Hitler gave number-one priority to the gasoline-burning V-1 "buzz bomb" and to the 46-foot-tall V-2 ballistic missile.

The V-2 rocket proved to be just the kind of weapon that Hitler was after. It was capable of reaching speeds of up to 3,600 miles per hour, covering as much as 200 miles. Carrying as much as a ton of explosives, the V-2 was impossible to stop once it was on its way to a target.

Even though the V-2 was designed to kill enemies in war, its use was by no means limited to such destructive purposes. It might also be used someday for transportation—to carry mail and also, possibly, human beings. Most important, it had the potential for reaching outer space.

Eventually, even with the breakthrough at Peenemünde, the German army began to crumble. By early 1945, both U.S. and

Wernher von Braun with rocket model

Soviet forces were converging on Germany. Wernher von Braun had decided to lead hundreds of top German scientists away from the Soviet army, advancing from the east, and they surrendered themselves as well as their vast knowledge in rocketry to the Americans in nearby Bavaria, to the west. Von Braun was immediately hired by the U.S. Army and sent to Fort Bliss, Texas, to continue his experiments on the V-2 and to explain how it worked to American scientists.

In the aftermath of World War II, the rivalry between the United States and the Soviet Union grew tense as each of the two countries fought to extend its power and influence. This competition was not only military but also scientific. Even so, the

A model of Sputnik, in the National Air and Space Museum in Washington D.C.

next war, this time in Korea, again demanded the services of the rocket scientists. Von Braun and his men were commissioned to develop a long-range nuclear missile. The two nuclear bombs that devastated Hiroshima and Nagasaki five years earlier were not rockets; they were simply dropped from the sky. The new rocket weapons could be launched from hundreds of miles away. Again, however, discoveries made during war had their consequences for peacetime. A similar rocket to those used in long-range nuclear missiles would be used in America's first unmanned spacecraft, Explorer I, in 1958.

The Soviet rocket program was pushing ahead as well, and at a pace somewhat faster than that of the American program. Between the years 1950 and 1955, many scientists began to explore the possibility of sending an artificial satellite into orbit around the earth. In 1951, the British Interplanetary Society sponsored a discussion in London entitled "The Artificial Satellite." Then, in 1957, the world scientific community decided to put together the greatest worldwide research effort ever. Thus was born the International Geophysical Year of 1957-58.

The head of the U.S. delegation was a man named Dr. Joseph Kaplan. Kaplan recommended that a satellite be launched as part of the American contribution to the program. The White House agreed with Dr. Kaplan, but what nobody noticed was that the Soviet Union had made a similar announcement three and a half months before.

The Soviet Union launched the first manmade satellite on October 4, 1957. The entire world, especially the United States, stood still in shock. Called Sputnik I, the 23-inch aluminum ball weighed 184 pounds and reached a height of 560 miles above the earth at its highest point as it orbited the planet. About an hour and a half after launch, the tiny ball had circled the globe, crossing over Russia for a second time. As it did, radio hobbyists around the world tuned in to hear the faint "beep-beep" signal coming from the first artificial object in outer space. What amazed the experts most about Sputnik was that it was shot into space at such a perfect angle, speed, and altitude, heading in such a way as to fly parallel to the earth's surface, that the force of the rocket balanced exactly with the gravity pulling it back to Earth. It is this principle that allows rockets and satellites to remain in orbit.

The day after Sputnik's launch, a New York *Times* reporter wrote, "The creature who descended from a tree or crawled out of a cave a few thousand years ago is now on the eve of an incredible journey." It was true. The giant Soviet achievement proved that the dreams of such historically distant figures as Copernicus and Robert Goddard, Leonardo da Vinci and Konstantin Tsiolkovsky, Galileo and Wernher von Braun, could be realized. The success of Sputnik reflected the same spirit that von Braun had exhibited when he was asked what it would take to build a rocket to take someone to the moon. His answer was, simply, "The will to do it."

2

THE JOURNEY BEGINS

Even while the shock of Sputnik I was still fresh in American minds, the Soviet Union readied Sputnik II, a heavier satellite. Although it was similar to the first one, there was one important difference. Sputnik II was to carry the first living space traveler, a little black and white dog named Laika. At last the question that had been on people's minds for centuries might be answered: Could a living creature survive a trip out of Earth's atmosphere?

The 1,120-pound Sputnik II was successfuly launched exactly one month after Sputnik I. In addition to its living cargo, it carried a great deal of scientific equipment. After a week in space, the dog died of asphyxiation.

Now it was America's turn. Although the Soviet satellites had caught many by surprise, U.S. scientists had been developing their own satellite, Vanguard I, to be America's contribution to the International Geophysical Year. And they were counting on it to offset the stunning Russian successes.

A test launch was set for December 6, 1957—just two months after the launching of Sputnik I. The countdown began at one in the morning. As spectators gathered, they saw a coat of frost appear on the first stage of the rocket, signaling that the freezing-cold oxygen was being pumped into it. Hours later, at 11:44 A.M., the umbilical cords connecting the rocket to the tower dropped away. Flames shot out from beneath the rocket. Vanguard began to rise.

What followed was a humiliating moment. Vanguard, the United States' first attempt to send a satellite into outer space, rose to a whopping height of four feet,

Vanguard explodes on the launch pad

then fell back onto the launch pad and exploded into a ball of fire seventy feet high.

Nothing could have made the United States look more inept. After the Soviet Union had awed the world by its successful launch of a 1,120-pound satellite that had sustained the life of a dog for an entire week, the U.S. had failed to get a satellite weighing a mere 3.25 pounds more than a few feet off the ground. Scientists around the world suggested that it would take years for America's space program to catch up to the Soviet Union's.

Von Braun, by now a U.S. citizen and director of the Development Operations Division of the Army Ballistic Missile Agency in Huntsville, Alabama, later commented, "Overnight, it became popular to question . . . our public education system, our industrial strength, international policy, defense strategy and forces, the capability of our science and technology. Even the moral fiber of our people came under searching examination."

Even before the catastrophic first test of Vanguard, the U.S. government had instructed von Braun to begin work on a rocket of his own design. Now von Braun vowed that the U.S. would have a satellite in space within ninety days.

He kept his promise. On January 31, 1958, a pencil-shaped, 31-pound satellite, boosted by a Jupiter-C rocket designed by von Braun, was launched. It was called Explorer I. The lift-off went perfectly. Less than two months later, on March 17, 1958, a second satellite, Vanguard I, rose into

orbit. The U.S. space program was back on track.

With space exploration no longer a dream, President Dwight D. Eisenhower advised Congress to create a special federal agency to run the space program. It was called the National Aeronautics and Space Administration—NASA, for short. The U.S. space program had formerly depended entirely on the army for support; this new agency gave it complete independence.

Shortly after NASA began operating in October 1959, it was alloted not only the permanent services of von Braun and his Huntsville staff, but more than $100 million in research facilities. The Explorer and Vanguard satellites had been successfully launched. The next step was to send a human being into space. To accomplish this goal, NASA designed three different programs. First, there was Project Mercury, a series of missions that would put the first Americans into space and then into Earth's orbit. Next would come Project Gemini, in which a pair of astronauts would test NASA's capability for successful space walks and other maneuvers. Finally, the third series of missions, Project Apollo, would seek the ultimate goal: Americans on the moon.

The search began for the first brave American men who would go into space. Five hundred eight military test pilots volunteered and were given every imaginable kind of test—tests of skill, of strength, of endurance, of bravery. The field was narrowed to 110 pilots, then to 58, and finally to 32. These 32 were sent to a civilian clinic in Albuquerque, New Mexico, for additional medical tests. They were submerged in water to test their response to weightlessness; they were given seventeen separate vision tests. Only one of the thirty-two men failed.

The pilots who had qualified for the program were flown to the Aeromedical Laboratory at Wright-Patterson Air Force Base in Ohio. There they were tested to see how much heat, pressure, noise, vibration, acceleration, and zero-gravity they could stand. They ran on treadmills, blew up balloons till they could blow no more, and spent a week undergoing psychiatric tests. Finally, eighteen candidates were chosen. All of them were brave military test pilots, under forty years of age, in top condition both physically and mentally, and weighing 180 pounds or less. Most important, all of them were hard to scare.

Finally the list was pared down even further. Seven men emerged as America's first astronauts: Malcolm "Scott" Carpenter, Leroy "Gordon" Cooper, John Glenn, Virgil "Gus" Grissom, Walter "Wally" Shirra, Donald "Deke" Slayton, and Alan Shepard. The final choice had been made not so much for scientific reasons but because it was thought that these seven

seemed to get along well with one another and would make a good team. Each of them, however, was competitive and personally ambitious. And each of the seven astronauts hoped to be the first American in space. Their names were announced at a press conference, instantly making them America's newest heroes.

Now that the men had been selected, a spacecraft had to be built that could support one or more of them in space. This was a difficult task. NASA knew that von Braun's Redstone rocket could get the spacecraft and its passengers into space.

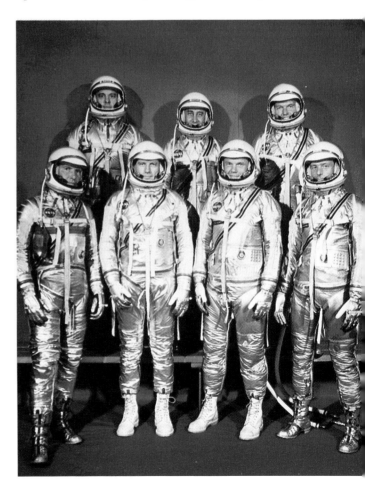

The seven astronauts of the Mercury program: Front row, left to right, Walter M. Schirra, Jr., Donald K. Slayton, John H. Glenn, Jr., and Scott Carpenter; back row, Alan B. Shepard, Jr., Virgil I. "Gus" Grissom and L. Gordon Cooper

But once there, the spacecraft would have to separate from the rocket and make its way back to Earth under its own power. What would protect the astronauts from the searing 3,000-degree heat they would encounter when the spacecraft reentered Earth's atmosphere?

A NASA engineer, Maxime Faget, had an idea: The spacecraft could be a funnel-shaped capsule that would sit on top of the rocket, its pointed end up. As it detached from the rocket in space, it would then roll over and fly flat end first. That end could be covered with a protective shield designed to withstand the dangerous heat of reentry.

While NASA engineers worked on the spacecraft, the astronauts began their intensive training. To test the ability to handle the unusual force of gravity in space, they took turns being strapped into a machine that swung them in circles until their muscles and bones were squeezed 16 times harder than normal. They were made to endure 135-degree heat for as long as two hours. And they rode a "Multiple Axis Space Test Facility," rotating 30 times a minute, to see how long it took them to control dizziness should the spacecraft begin to spin around in space.

Finally, the space capsule was ready for the astronauts to inspect, and they did not like what they saw. What bothered them was not that it was small and cramped but that there was only a periscope and two tiny portholes to see out of. Why couldn't it have a larger window? they asked. The engineers explained that a large window

might not withstand the pressure of space. Nonetheless, the astronauts were insistent. The engineers went back to their drafting boards, and the astronauts got their window.

The astronauts' second complaint had to do with safety. The capsule had seventy bolts on it that took more than an hour to tighten from the outside. Suppose they had to get out of the capsule quickly? A new hatch was designed, one with explosive bolts that could be detonated from inside with the push of a button.

The astronauts' final and most serious complaint about the spacecraft was that its control system entirely ignored their skills as ace test pilots. Since the craft was designed to be controlled by computerized boxes from the ground, their role, they felt, would not be too much different from that of the group of experimental monkeys and chimpanzees that would be preceding them into space. Finally, it was agreed that the control system would be modified so that if the computers failed, the astronauts would be able to take over the controls themselves.

Meanwhile, less than two months after their names had been announced to the world, the seven astronauts watched from the ground as a series of monkeys and chimps became America's pioneers in space. Although previous U.S. flights had taken up fruit flies and mice, the "astrochimps," so similar to humans, were far better test subjects. Not only could the reactions of their bodies be more easily monitored, they could even be taught to

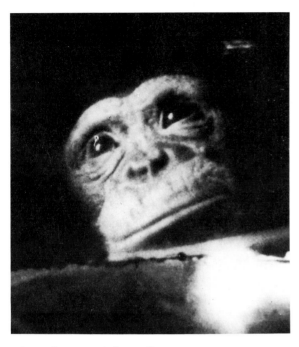

Ham the "astrochimp"

pull levers in the capsule on command. The most famous of these primates was a little fellow called Ham, who rode the Mercury Redstone 2 rocket, designed by von Braun—the same rocket that was to take the first man into orbit. Ham's successful flight led to the announcement that the date had been set for the first astronaut to go up—May 5, 1961. The country braced itself, ready to watch the first American—and the first human—ascend into space.

This was not to be, however. On April 12, less than one month before the scheduled U.S. flight, a Russian radio announcer broke the news: "The world's first spaceship, Vostok, has been launched. . . . The first space navigator is Soviet Citizen Pilot Major Yuri Alekseyevich Gagarin."

The Soviet flight had gone perfectly. Gagarin had become not only the first person in space but the first to orbit the earth. Soviet citizens were ecstatic. Safely back on earth, Gagarin appeared in public to lead a parade from Lenin's Tomb in Moscow's Red Square. Premier Nikita Krushchev even gave Gagarin a new title: First Hero Cosmonaut.

Though Americans were awed by the Soviet flight and admiring of the brave cosmonaut, they also felt frustrated. Once again, the USSR had upstaged the USA. Would it never end? As one member of Congress said upon hearing the news, "Everyone remembers Lindbergh, but who remembers the second man to fly the Atlantic?"

That second man in space was to be U.S. Navy lieutenant commander Alan Shepard. Chosen as the first American astronaut for a suborbital flight, he lay on his back the morning of May 5, 1961—exactly as scheduled—in a capsule high atop a Redstone rocket at Florida's Cape Canaveral. Awake since 1 A.M. and strapped in the capsule since 5:20 A.M., he waited as the hours passed, eager to get going. At one point he asked the ground crew, "Why don't you fix your little problems and light this candle?"

Finally, the Mercury-Redstone 2 blasted off, rising into the skies. At 116 miles high, Shepard radioed back to Mission Control. "What a beautiful view!" The flight proceeded without a hitch. Fifteen minutes later Shepard and his capsule, Freedom 7, splashed down in the ocean and were picked up just four minutes later by helicopter.

Alan Shepard being hauled out of the ocean after the splash-down of the Mercury program's Freedom 7

On July 21, 1961, just over two months after Shepard's historic flight, a second, rather hair-raising suborbital flight took place with astronaut Gus Grissom aboard. The liftoff and flight went perfectly, as before. Problems began after Grissom's module, the Liberty Bell 7, had landed in the ocean as planned and Grissom was waiting to be picked up by helicopter. "I was just lying there, minding my own business, when I heard a dull thud," he recalled later.

Suddenly the hatch blew open and water poured into the capsule. Seconds later Grissom was outside the capsule, try-

ing to stay afloat in the ocean. But then he realized that his space suit was leaking air from an oxygen valve he had forgotten to close in all the excitement. Just in the nick of time, however, he managed to grab hold of a line that had been dropped from one of the helicopters hovering overhead. Pulled to safety, he and the helicopter crews watched helplessly as the Liberty Bell, which had been filling with water, sank to 2,800 fathoms below the sea.

Despite the lost capsule, NASA felt ready to go on to the next step—an orbital flight—as scheduled. For this historic mission, which would entail orbiting the earth three times. Colonel John Glenn (later Senator Glenn of Ohio) was named as pilot. Because the Soviet Union's second cosmonaut, Gherman Titov, had suffered from dizziness in space, Glenn spent part of his preflight training doing a series of head and body exercises designed to prevent the same from happening to him.

Glenn's flight in Friendship 7 was a great breakthrough. During his first orbit high above the earth, he saw the Canary Islands off Africa, then the Nigerian coast. Soon he was over the Indian Ocean, and he viewed a brilliant sunset from 100 miles above the earth. Although the sun didn't set as quickly as he'd imagined it might, gradually the vivid orange and blues surrounding the sun began to fade as he passed over India and neared Australia. Over the orbital-tracking station in Muchea, Australia, he struck up ground-to-space contact with fellow astronaut Gordon Cooper, who was stationed there for the flight. Excitedly, he told Cooper he could see the lights of the city and the clouds reflecting the moonlight on the horizon. "That sure was a short day. That was about the shortest day I've ever run into," he said. Moments later he was over the Pacific island of Canton, watching the sunrise.

Glenn continued orbiting until he'd seen three sets of sunrises and sunsets—all in five hours. While nearing the California coast for the third time, the first retro-rocket came on, followed seconds later by a second and third, braking the orbit of the ship in preparation for reentry and splashdown. Glenn then maneuvered Friendship into the necessary 19-degree angle for the dangerous descent into the atmosphere.

John Glenn

Suddenly there was trouble. Mission Control began getting readings that the all-important heat shield on the capsule was loose. If it fell off, Glenn would burn up as he returned to Earth. Aboard Friendship, Glenn began to hear noises; it sounded as if "small things [were] brushing against the capsule," he reported. Outside, flames engulfed the spaceship. "That's a real fireball outside, boy," Glenn radioed, with just a hint of nervousness in his voice. Luckily, however, the heat shield did hold up as the spaceship plunged through the earth's atmosphere. By the time the capsule's parachutes were activated, the scare was over. John Glenn and Friendship were safe.

There were three more Mercury orbital flights between May 1962 and May 1963. The first was Scott Carpenter's, aboard the Aurora 7. This flight, primarily a scientific expedition, ran into some problems—excessive fuel consumption, overheating in both the cabin and Carpenter's space suit, and a fault in the automatic controls that caused Aurora to splash down 250 miles off target. But Carpenter was all right. Next to go up was Wally Shirra, in Sigma 7. His flight was much less eventful than Carpenter's, his main task being to take as many photographs as possible. Finally, on May 15, 1963, Gordon Cooper went up in the Faith 7. Cooper spent 34 hours in space, orbiting the earth 22 times. It was a flawless flight, the perfect end to the Mercury program.

The next step in NASA's path to the moon had been determined by world politics. In May 1961, just twenty days after

President John F. Kennedy

Alan Shepard's historic suborbital flight, President John F. Kennedy had made a stirring vow to the American people that was to change the course of the nation's space program:

> "I believe this nation should commit itself to achieving the goal before this decade is out of landing a man on the moon and returning him safely to the earth. No single space project in this period will be more impressive to mankind, or more important for the long-range exploration of space; and none will be so difficult or expensive to accomplish."

Quickly, NASA engineers began rethinking their plans for the moon mission. The original plan had called for a direct ascent, with the capsule that had been launched from Earth going all the way to the moon and back again. On its return, two huge on-board rockets would have to be fired to give it the necessary power to get it back to Earth. But that was the problem. These powerful rockets were still on the drawing boards; they wouldn't be ready till the end of the decade. There had been a second plan, but it didn't look much better. This one called for two separate launches— one spacecraft and one vehicle for extra fuel. The two would meet in Earth orbit, transfer fuel, and then the spacecraft would proceed on to the moon and eventually return to Earth. Yet, as NASA engineers thought about it, they realized this would require too much fuel.

A third idea solved the problem. In this plan, two attached spacecraft would be launched into orbit around the moon. The smaller one would separate, make the landing, then take off from the moon and rejoin the other craft. The landing module would then be jettisoned and the larger spacecraft, now much lighter, would return to Earth alone. Under this plan only half as much fuel would be required, and only one launch. Still, the rendezvous and docking in space would be a tricky matter.

That is where Project Gemini came in. Many questions had to be answered before this kind of moon landing could be attempted. Could astronauts survive an entire week or more of weightlessness? Could they carefully control their reentry into Earth's atmosphere? And what about their ability to survive while space-walking outside the craft?

Work began on the Gemini spacecraft. Since the mission called for two astronauts, the craft had to be larger. When completed, it weighed almost twice what the Mercury craft had weighed and was a foot longer and wider. "It looks like a Mercury capsule that threw the diet rules away," said Gus Grissom.

While work was going forward on the Gemini capsule, NASA announced the addition of twenty-three astronauts in the fall of 1962. Gus Grissom and John Young, one of the new additions, were then named as the two space travelers for the first Gemini mission. Because he'd watched the first capsule he'd been in sink into the ocean, Grissom named his new Gemini capsule *Molly Brown*, after the Broadway musical *The Unsinkable Molly Brown*. NASA officials didn't much like the name, but they gave in after Grissom mentioned that his second choice was *Titanic*.

Two years passed between the last Mercury flight and *Molly Brown*'s launch on

Ed White walks in space

March 23, 1965. The flight proved worth the wait. Although the Soviet cosmonauts had executed a successful space walk just five days earlier, the two Americans performed maneuvers with their ship that had never been done before. During the first orbit, Grissom used the spacecraft's sixteen small thrusters to change the course of orbit to an almost circular loop. This was a major achievement in space travel because it proved that the astronauts could really control their craft—a condition that would be critical for any future moon landing. Splash-down fell fifty miles short of the recovery ship, however. During the half hour that Grissom and Young waited for the helicopters to pick them up, Grissom refused to open any hatches. "If *Molly Brown* had sunk out there," he said later, "I'd have jumped right off that carrier."

During the next three years, Americans were treated to the drama of one Gemini flight after another, each more ambitious than the one before. During the second Gemini flight, astronaut Edward White became the first American to walk in space, with a twenty-four-foot line connecting him to the spacecraft. "I can sit out here and see the whole California coast," he told fellow space traveler James McDivitt. Indeed, the experience was such a thrill that it took a good deal of coaxing to get him back inside the craft.

In December 1965, the first attempt at rendezvousing in space took place. First to go up was Gemini 7, with astronauts Wally Shirra and Thomas Stafford aboard. Eight days later, Gemini 6 was launched, carrying Frank Borman and James Lovell. Just six hours after its launch, the second spacecraft was within 120 feet of the first. For the next five hours, the two ships flew in formation, at one point only one foot away from each other. After both pairs of astronauts napped, Gemini 6 went home. Gemini 7 stayed on, setting a new endurance record of 14 days in space.

Both Gemini 8 and 9 had problems. After Gemini 8 had successfully rendezvoused with an empty target vehicle, the two spacecraft suddenly began to spin end over end, faster and faster. Neil Armstrong tried to stop the spinning by using Gemini 8's maneuvering thrusters, but that didn't work. Even after the undocking from the target vehicle, the spinning continued. It wasn't until Armstrong had deactivated the manuevering system and fired the reentry rockets that the spinning finally stopped. But the remainder of the trip had to be aborted to be sure they had enough fuel for reentry.

Gemini 9's mission called for rendezvousing and docking with a target vehicle and for astronaut Eugene Cernan to leave the capsule wearing a Buck Rogers-style backpack. But a shroud designed to protect the target vehicle's docking cone had not detached during the launch, so docking was impossible. Besides that, Cernan had trouble with the backpack. Putting it on required such effort that his suit's cooling system overloaded. By the time he managed to put it on, Mission Control was ready to recall him. Although he spent a record of 2 hours and 8 minutes outside the

Gemini 7 as photographed through the hatch of the Gemini 6

spacecraft, that was still 39 minutes short of NASA's goal.

The next two flights, however, were dramatic achievements. Gemini 10 actually managed to rendezvous with two target vehicles in space, and one of the astronauts aboard, Mike Collins, became the first man to take two spacewalks on the same mission. Two months later, in September 1966, Gemini 11 set a record for reaching the greatest distance from Earth: 855 miles! The astronauts aboard also succeeded in linking two craft together with a 100-foot tether.

Even though Gemini 10 and 11 had been successes, questions were raised about how truly successful the Gemini program had been in terms of spacewalking. Both Collins and Richard Gordon, aboard Gemini 11, had been totally

exhausted, with Collins momentarily getting tangled in his umbilical cord. If Americans were ever to walk on the moon, this problem would have to be solved.

So, for Gemini 12, NASA not only made some equipment changes but also trained astronaut Buzz Aldrin specifically for the pressure of weightless movement in outer space. The extra care paid off. On this final Gemini space walk, Aldrin spent 5 hours 20 minutes outside the ship, reporting no fatigue whatsoever.

When Aldrin and James Lovell splashed down aboard Gemini 12 on November 15, 1966, it was clear that Project Gemini's entire series of missions had been a success. While the Soviets had remained on the ground, 16 Americans had spent 1,940 hours in space at a cost of $1.1 billion. It was time for the third and final phase of

Gus Grissom, Ed White and Roger Chaffee inside the practice module

NASA's quest for the moon—Project Apollo—to begin.

Even as the Gemini Project was at its busiest, NASA technicians had been hard at work on Apollo. Although Gemini proved that U.S. spacecraft could maneuver, rendezvous, and dock with other craft, scientists had to find out exactly what the astronauts would encounter on the moon. Since 1964, NASA had been sending up a series of unoccupied lunar probes to photograph the moon's surface. The first of these—the Ranger probes—were able to take pictures and send them back to Earth, but then they would crash into the moon, unable to return home. The later Surveyor probes, which did not crash, could continue photographing and sending the results back. Along with the Soviet Luna probes, Surveyor eased the fears of many scientists that the moon's surface was so dusty that any landing craft would be buried in the dust.

Meanwhile, NASA technicians were designing the rocket and spacecraft to be used for the moon landing. The Saturn V was the most powerful rocket ever built, standing more than 360 feet high. Each of its five first-stage rockets was the size of a 2½-ton truck. Together they supplied the same amount of energy as 86 Hoover Dams.

The first Apollo flight, though, did not use the Saturn V rocket. It was set for February 1967, just three months after Gemini 12. The three crew members chosen for the flight were veterans Gus Grissom and Ed White along with rookie Roger Chaffee, who, at thirty-one, was the youngest astronaut to date.

Three weeks before the launch, the three men were practicing a countdown and simulated launch. All afternoon there had been problems, so many that Gus Grissom had hung a lemon on the side of the spacecraft. Then, at 6:31 P.M., one of the astronauts, possibly Chaffee, said calmly on the intercom, "Fire. I smell fire."

A television camera that was focused on the capsule's hatch suddenly recorded an enormous flash. A loud scream was heard, then silence. Seconds later the capsule exploded with fire. By the time it could be put out, Gus Grissom, Ed White, and Roger Chaffee were dead.

The entire nation was stunned. For a

The practice module after the fire

while it looked as if the whole Apollo program might be scrapped. Hundreds of government investigators sifted through the burned remains of the spacecraft, looking for the cause of the tragedy. Although they were never able to pinpoint the cause exactly, investigators concluded that the fire had started near a bundle of wires in front of Grissom's couch. Apparently, one of these wires had sparked, igniting flammable materials that burned explosively in Apollo 1's highly pressurized atmosphere of pure oxygen. A 3,000 page report on the tragedy led to many changes in the Apollo spacecraft. But NASA had to abandon its original schedule for missions while each of the three major parts of the system—the command module, the lunar module, and the Saturn V rocket booster—was tested, repaired, and retested.

After several test flights with no crew, NASA was once again, after an interval of twenty months, ready to send astronauts into space. Apollo 7, sent on a mission to test the command module in Earth orbit, carried Wally Shirra, Don Eisele, and Walt Cunningham. During the flight, the astronauts proved the maneuverability of the command module by making a rendezvous without radar with the second stage of their Saturn booster. But what pleased the three men most about the flight was the roominess on board. Said Cunningham, "We actually had living quarters, not just a place to sit."

Apollo 8 featured the first use of the Saturn V, the only rocket yet built that was powerful enough to propel a spacecraft beyond Earth orbit. Aboard were Frank Borman, Jim Lovell, and William Anders. Their mission: to circle the moon. After a trip of 3 days, Apollo 8 came within sixty miles of the moon on Christmas Eve, 1968. The next morning, Christmas Day, Borman fired the command module's engine to escape the moon's orbit for the trip home. Clearly, America had taken a great step closer to landing on the moon.

One important test still had to be done. On Apollo 9, astronauts Jim McDivitt, Dave Scott, and Russell Schweickart actually flew the lunar module one hundred miles away from the command module before coming back for a successful docking.

Now the only thing left was the dress rehearsal—Apollo 10, carrying Thomas Stafford, Gene Cernan, and John Young. This mission did everything that Apollo 11, the actual moon flight, would do except land. When it splashed down safely, NASA was completely satisfied that all its preparations were complete.

In fact, the astronauts decided, NASA might have been just a bit too prepared. For this last flight they had invested $5,000 on something called "scientific experiment Sugar Hotel Alpha Victor Echo," or SHAVE, which was meant to prevent the astronauts' excess facial hair from clogging up the instruments on board. But the

Expended second stage Saturn V rocket jettisoned approximately three minutes after liftoff

men found that a drugstore safety razor and brushless lather did the job just as well. Anyway, the problem of shaving in space wasn't going to hold back the United States from achieving one of humankind's greatest goals.

As far as the moon landing was concerned, all systems were go.

3

MISSION ACCOMPLISHED

When Neil Armstrong, Buzz Aldrin, and Mike Collins heard that they had been chosen for the Apollo 11 flight, Armstrong suspected that it would not be the flight that would actually land on the moon. There were still too many unknowns. At that time, several months before the scheduled launch, it hadn't yet been proved that the ground crew could communicate with two vehicles at once. Nor had the lunar module been properly tested.

Although the success of Apollo 9 and 10 eased Armstrong's worries, scientific questions about the nature of the moon itself remained. How was it created? How had it evolved? Part of Apollo 11's mission would be to investigate such questions. But, of course, the primary mission would be just to get there and get back. That's why the most important factor in the

Astronauts of the Apollo 11 mission

choice of the three astronauts had been that all of them were known for their superior piloting abilities.

Neil Armstrong, the mission commander, had wanted to fly ever since he could remember. As a little boy in Wapakoneta, Ohio, he had often dreamed that he could hover above the ground just by holding his breath. By fourteen, he was saving every penny he made working at the local pharmacy for flying lessons. The day he turned sixteen, August 5, 1946, he got his flying license—well before he had a license to drive a car.

After studying aeronautical engineering at Purdue University, Armstrong flew Panther jets from the carrier *Essex* during the Korean War. He flew seventy-eight combat missions and won three Air Medals. When his service in Korea ended, he became a civilian pilot at Edwards Air Force Base in California. In his seven years there,

Neil Armstrong   *"Buzz" Aldrin*   *Michael Collins*

he survived several near-fatal plane accidents, including one in the experimental, rocket-powered X-15, which had reached speeds of up to 4,000 miles per hour. He became an astronaut for NASA at about the time that Wally Shirra was taking the next-to-last Mercury flight. With his training completed, Armstrong went into space aboard Gemini 8, a particularly hair-raising flight and the only one of NASA's missions that had to be aborted in space.

Clearly, though, Neil Armstrong was a man with a talent for staying alive. He was not only one of the best pilots around but one of the most fearless—exactly the kind of person needed to be the first human being to explore the moon.

Buzz Aldrin, the astronaut chosen to be second on the moon, was of equally tough stuff. He was known for having the best scientific mind that NASA had sent into space. A graduate of the Massachusetts Institute of Technology (MIT), he was also the first astronaut with a doctoral degree.

Edwin "Buzz" Aldrin was the son of a distinguished pilot of the 1920s and '30s. His father had known Orville Wright and had been a student of Robert Goddard's. Indeed, the elder Aldrin had been the one who introduced Goddard to Charles Lindbergh in 1929, a meeting that eventually led to the generous grant Goddard received from the Guggenheim Foundation.

Buzz Aldrin attended the U.S. Military Academy at West Point in the 1950s. After graduation, he spent a year in Texas getting his Air Force wings and then was assigned to Korea. During his four years there, he flew sixty-six combat missions and was awarded numerous decorations. Back at home, he became an Air Force instructor in Nevada for a while, and later, in Germany, flew F-100s. But Aldrin began to feel a need to do something more challenging. In 1962 he applied to be one of the new astronauts. Because he'd had no experience as a test pilot, he was turned down. Meanwhile, at MIT, he was in the midst of work on "some original ideas about piloting problems and rendezvous."

Now NASA was interested. Buzz Aldrin not only became an astronaut but went on to demonstrate his work at MIT during his successful space walk on Gemini 12. Clearly it had been his determination and constant quest for new challenges that had taken him into space. And now it was precisely those qualities that were leading him to the moon.

The third member of the Apollo 11 flight crew, Mike Collins, almost had to leave the space program when doctors found that a piece of bone pressing down on his spinal cord was causing him to lose feeling in parts of his body. With a worsening of his reflexes, he had even begun falling down occasionally. Doctors gave him two choices: Either he could have a relatively safe operation that would definitely relieve his problem but not cure it or he could undergo a very dangerous surgical procedure that might clear up the problem for good. For Collins, the choice was easy. He was too much of a flyer to contemplate a future on the ground. He took the second of the choices, and the operation was a success. Although he missed the chance to orbit the earth on Apollo 8 because he was recuperating, at least he knew that his career as an astronaut would not come to an early end.

Collins had grown up in a military family, moving from one army base to the next. Since his father was a major general and his uncle a general, it was natural for him to enter West Point. Upon graduation, he spent four years in Europe training in fighter planes. Back in this country, he joined the test-pilot program at Edwards Air Force Base, where he flew X-15s and other experimental planes. Then, like so many other test pilots at Edwards, including Armstrong, he turned to outer space, which led him to a mission aboard Gemini 10 in 1966. Why did Collins, like so many of the other astronauts, keep pushing himself on to new frontiers? "I think man has always gone where he could; he's always been an explorer," Collins said in a *Life* magazine interview just before his Apollo 11 flight. "I really think the key is that man has gone where he could and must continue. He would lose something terribly important by having that option and not taking it."

The world stood by in July of 1969, waiting for Collins and Neil Armstrong and Buzz Aldrin to go on their historic voyage to the moon. As Wernher von Braun commented the night before the launch, "What we will have attained when Neil Armstrong steps down upon the moon is a completely new step in the evolution of man. For the first time, life will leave its planetary cradle, and the ultimate destiny of man will no longer be confined to these familiar continents that we have known so long."

Aware of the symbolic importance of their mission for people the world over, the three astronauts had taken particular care in naming their two spacecraft. "Eagle" was chosen for the lunar module, to reflect America's national symbol, and "Columbia," the name of the command module, was taken from Jules Verne's spaceship, *Columbiad*, in his book *From the Earth to the Moon*. Most important, Columbia was chosen because it related to another great discovery—the discovery of the Americas by Christopher Columbus.

U.S. Highway 1, leading to Cape Canaveral, was lined with trailers, vans, and tents of every kind on the morning of the launch. People were everywhere, watching from shore ten miles away and rocking back and forth in boats of all sizes that dotted the surrounding waters. As the count-

Apollo/Saturn second stage separation

down reached its final moments, millions of television sets were tuned in around the world as people watched with anticipation.

Then it happened. As far as ten miles from the launch site, spectators saw a burst of orange flame with a stream of grayish smoke billowing behind it. Upward and upward into the sky the rocket pushed. The flames shooting from the rocket's base burned so brightly that those watching had to avert their eyes. Thirty seconds after launch, Apollo 11 was invisible to people on the ground. Less than a minute after take-off, it was traveling faster than the speed of sound.

Eleven minutes after launch, Apollo 11 was in Earth orbit. Very carefully, the astronauts began to move about, making sure they did not feel any of the nausea reported by previous crews. As Columbia's pilot, Collins had the most chores to complete at this stage. Armstrong and Aldrin spent most of their time adjusting to weightlessness and catching various objects that were floating around the cabin. After one and a half orbits, the boosters on the Saturn V rocket fired, sending the ship out of Earth orbit and on toward the moon.

Now Collins had to separate the Columbia from the rocket's third stage, turning it around and connecting it to the Eagle, which was stored in the third stage. This was a critical procedure. If it didn't work, the astronauts would have to return to Earth. But the four panels enclosing the Eagle automatically came off and drifted away as planned. Still another, even more

dangerous possibility lay ahead. In the docking operations there was a chance of a midair collision between the Eagle and the Columbia. Knowing that if that happened the Columbia's cabin would immediately decompress, the three men remained in their space suits. The docking, however, went perfectly. With the nose of the Columbia now connected to the top of the Eagle, the Apollo 11 continued on its way.

Since the docking was behind them, the astronauts took off their bulky space suits and changed into more comfortable jumpsuits. But they still had important work to do. In order to prevent excessive heat from building up outside of Apollo 11, and to ensure continuing communications with Mission Control, they had to keep the spacecraft in constant rotation around its own axis. After completing some complicated maneuvers, the men took advantage of the view this spin gave them. As the Apollo 11 sped along through space at a rate of 25,000 miles an hour, every two minutes the sun, Earth, and the moon would appear in their windows at the same time. Soon, the planet Earth became so small that according to Aldrin, he could block it out of the universe simply by holding his thumb up to it.

As far from home as they were, they still had to observe their old, earthly eating and sleeping habits. Dinner that first night was shrimp cocktail, soup and crackers, and some cheese and meat spreads. Then, fourteen hours after lift-off, it was time for some sleep. For this, they took turns.

While one man stayed awake to take any calls from Mission Control, the other two stretched out in mesh sleeping bags located under two of the cabin's couches.

Bright and early the next morning, Mission Control woke them up with a cheery "Good Morning" and a piece of news. The Soviet Union had sent up an unmanned probe to the moon, they were told, and the White House was worried that Apollo 11 might crash into it. The astronauts were more amused than afraid. The chance of the Soviet satellite's hitting them, said Collins later, "was about equivalent to my high-school football team beating the Miami Dolphins."

The second day of the mission was, as expected, more leisurely. After some morning housekeeping chores and lunch, they hooked up a television camera in order to record what was happening for the millions of people, now 100,000 miles away, back on Earth.

By the morning of the third day, Apollo 11 entered the orbit of the moon. Now there had to be a complete check of the

Eagle. Armstrong and Aldrin crawled inside to begin powering it up and to complete a long list of final preparations. Although the lunar module was then ready to go, the actual landing would not take place till the next day.

Finally, the big day arrived. Arising well ahead of schedule, the three men donned their full space suits, then went to work. Armstrong and Aldrin stepped into the Eagle, closed the hatch behind them, deployed the module's spiderlike landing gear, and undocked it from the Columbia.

At a speed of 1,200 feet per second, they began to descend. Armstrong and Aldrin watched tensely as the radar recorded the descent toward the moon's Sea of Tranquillity. Suddenly, at about 30,000 feet above the moon's surface, the Eagle developed computer problems. Because of their training, the two knew that the computer would flash a yellow alarm light and

The lunar module Eagle descends to the surface of the moon

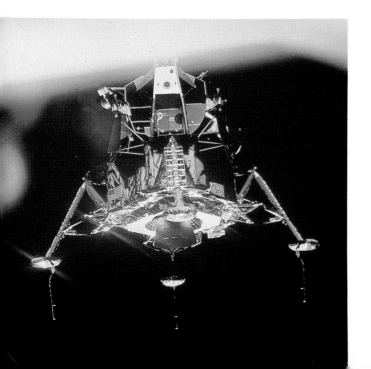

a number, not only signaling that there was a problem but also indicating what the problem was. But never in all their training sessions had either of the astronauts encountered a situation similar to what the computer was registering now. After quickly analyzing the situation, Mission Control advised them to ignore it and continue the descent.

At 3,000 feet, the two men noticed an area of the moon now called West Crater, which was about 200 yards in diameter and filled with rocks, some more than five and even ten feet across. Deciding to overfly this area, they headed for a flat surface between two boulder fields. As the rate of descent slowed to only 3 feet per second, the feelers on the Eagle's landing pads touched down.

"Contact light! Okay. Engine stop," said Aldrin.

Next came the voice of Commander Armstrong. "Houston. Tranquillity Base here. The Eagle has landed."

Back on Earth, it was 4:17 P.M., Eastern Daylight Time, July 20, 1969. At the Manned Spacecraft Center, Thomas O. Paine, an engineer who was the administrator of NASA, stood on stage a few minutes later, his hands clasped, like those of a preacher about to give a sermon. He had just phoned President Nixon at the White House, telling him, "The Eagle has landed in the Sea of Tranquillity, and our astronauts are safe and looking forward to starting the exploration of the moon."

Nearly six hours later, after both astronauts had taken short naps and checked

their equipment, Armstrong opened the Eagle's hatch. Squeezing through with the huge backpack that held his portable life-support system, he began his climb down the nine-step ladder. Briefly, he paused on the second step to deploy the camera at the side of the lunar module. Then, as he planted his left foot on the moon at 10:56 P.M., he spoke those stirring first words that would go down in history.

After Aldrin had joined Armstrong, feeling, as he put it, "buoyant and full of goose pimples," the two men spent two hours working on the moon's surface. They set up scientific instruments, collected rock samples, and planted an American flag with an arm on it that made it look as if it were waving in the windless moon air. Nor did they forget their brave fellow astronauts who had paved the way to this moment. Before getting back into the Eagle for the return trip to Earth, they left behind a small memorial to the crew of the ill-fated Apollo 1—a shoulder patch that was to have been worn by Grissom, White, and Chaffee. Then, after a short nap in Eagle's hammocks, the men turned on the ascent engine that provided 3,500 pounds of thrust for the lunar module's return to the command module.

The flight home with Mike Collins aboard the command module, Columbia, was smooth and uneventful. Back on their own planet again, the three astronauts had to spend their first seventeen days in quarantine, isolated from the public, to make sure they had not brought back unknown bacteria or other microbes from the moon.

The Apollo 11 astronauts speaking to President Nixon from the post-flight quarantine unit

The moon landings did not end with Apollo 11, however. The program continued for three more years, with ten additional manned flights that allowed twenty-four Americans to step out onto the barren, unexplored landscape of another world. These astronauts not only brought back to Earth nearly half a ton of lunar rock and soil, some of which proved to be 4.6 billion years old; they also brought home a new appreciation of the fragility of Earth and a sense of awe for the beauty of their newly expanded home, the universe.

4

SPACE EXPLORATION CONTINUES

Five months after the moon landing, NASA was hard at work on a new project, which marked the beginning of a new era in American space exploration. NASA was preparing to launch Skylab, a 118-foot-long orbiting space station.

Skylab was designed for several purposes: to enrich our knowledge of space, to study the effects of weightlessness on living beings, to find ways of making materials in the zero-gravity of space, and to investigate means of observing and monitoring Earth's ability to support the needs of all its people. Above all, Skylab was meant to test the capabilities and limitations of human scientists working in space.

Anticipation was high on May 14, 1973, when a Saturn V rocket lifted the

Skylab (note orange rectangular sunshade and missing solar panels)

unoccupied Skylab into orbit. But just one minute after lift-off, the space station was in serious trouble. Its thermal shield broke off, ripping off one of the two solar-powered wings folded against its sides. The other wing was pinned shut by a small piece of the broken shield. Without the thermal shield, Skylab was being toasted at a temperature of 325°F on the outside, 165°F on the inside. If the heat continued, food, film, and medicine would be spoiled. The heat even threatened to melt the ship's foam insulation, thus releasing deadly amounts of gas.

Mission Control sprang into action, firing Skylab's thrusters by remote control to tilt the burnt area away from the sun. This dropped the inside temperature to 130°F. Next, engineers designed a 22-by-24-foot rectangular umbrella. The first human occupants of the space station would have the task of installing the umbrella.

Eleven days after Skylab's launch, astronauts Pete Conrad, Joe Kerwin, and Paul Weitz blasted off to rendezvous with Skylab, armed with tools to fix the crippled ship. After seven and a half hours, pilot Conrad docked the Apollo command module nose to nose with Skylab. The crew then spent twenty-eight days in space, a new record. During that time, they patched up the damage by erecting the umbrella sunshade and unjamming the other solar wing. With the wing deployed, full power was restored, causing Pete Conrad to report, "Looks like we can turn on more lights and stop living like moles."

With the repairs completed, the crew now had time to perform some of their previously assigned chores. All three rode an exercise bike for half an hour each day and also tested each other for disorientation. Kerwin wore a caplike monitoring device so that NASA could see if his dreams were affected by living in space.

NASA scientists were pleased with the results of the mission. The crew, they said, had changed people's knowledge of how humans could survive in space. The men, too, had changed. After four weeks without the pressure of gravity, each man's spinal column had stretched. They had all grown one inch taller.

A month after the first Skylab crew had returned, a second crew went up, consisting of astronauts Alan Bean, Owen Garriott, and Jack Lousma. Again, there was an initial scare. One of the rocket clusters boosting the Apollo spacecraft sprang a leak. No sooner had it been fixed than another appeared. Wondering whether to abort the mission, Mission Control then learned that the two leaks were unrelated. The Skylab crew was told to continue on its way.

Problems were not over, though. The first three days after reaching Skylab, all three men were "spacesick," so much so that they could barely move. Finally recovering, they began their experiments. Nineteen of these had been suggested by the winners of a nationwide high-school science contest, most of whom wanted to see how nonhumans functioned in space. Mice, minnows, and fruit flies had been taken aloft. Although the mice and the fruit flies died, a pair of minnows survived. Swimming in small circles at first, the minnows eventually adapted to their new environment. When their offspring were born, they were completely normal, as if, said Garriott, "they'd adapted while still in the egg."

The astronauts had clearly adapted, too. After ten days Lousma and Garriott installed a second sunshade, causing Skylab's inside temperature to drop to the comfortable low 70s. When they splashed down on Earth again, they had logged a new record: 59 days in space, in which they'd traveled an amazing 24,500,000 miles.

This record did not last long. At 9:01 A.M. on November 16, 1973, rookie astronauts Gerald Carr, Edward Gibson, and William Pogue blasted off on the final Skylab mission, one that lasted twelve weeks. This was the busiest Skylab crew of

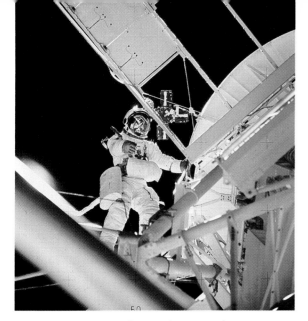

Owen K. Garriot, during his record seven-hour space walk

all. First, they were assigned a heavy workload, and because the previous crews had misplaced thousands of the tools on board Skylab, the work was more difficult. Thus, minor disagreements ensued between the crew and Mission Control over the jam-packed schedule. Still, much was accomplished. The men watched the seasons change in two hemispheres. They saw wheat ripening in Argentina, a volcano smoldering in Japan, bright lights blinking on in the Eastern United States.

On Christmas Day, Pogue and Carr took a record-breaking seven-hour space walk to take pictures of the comet Kohoutek. These were not the most exciting photographs on the mission, however. That honor went to Gibson, who caught the birth and life of a solar flare on film—a study having important implications for the control of nuclear fusion as an energy source on Earth.

With the completion of this third Skylab mission, the Skylab project ended. It

had been a big success. More than 180,000 pictures had been taken of the sun alone, vastly increasing people's knowledge of that star. Also, as Joe Kerwin said upon his return, space had been "kind to people." Skylab had proved that it was possible to live and work for an extended time in space.

Meanwhile, with the astronauts back on Earth, Skylab drifted alone and empty in space. Scientists predicted that it would remain in orbit until the early 1980s, when it would be dragged down into the dense layers of Earth's atmosphere and plummet downward, showering some countries, perhaps, with chunks of fiery metal. To prevent this red-hot rainfall, NASA hoped that the first reusable spacecraft, the Space Shuttles, would be ready for operation so

Solar flare, photographed by Edward Gibson

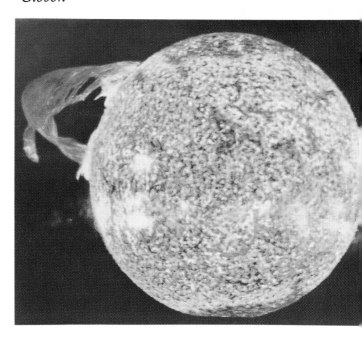

that one of them could nose Skylab into a higher, safer orbit. Unfortunately, however, the Shuttle was delayed. In addition, especially violent activity from the sun began expanding Earth's atmosphere, increasing the pressure on Skylab and causing its orbit to decay faster than expected. By early 1979, it was clear that Skylab was falling. Where it would land no one knew. Luckily, the doomed craft fell on an unpopulated part of western Australia and into the Indian Ocean, giving people in those areas a sudden, dazzling display of fireworks from above.

Actually, spacecraft have fallen to Earth more often than most people realize. In 1978, a nuclear-powered Soviet satellite fell on a deserted area of northern Canada, creating fears of a radiation hazard. And this was not an isolated accident. Soviet scientists often let their Salyut space stations "undergo orbital decay" when missions end. But since Salyuts are only a quarter the size of Skylabs, the risks are also smaller.

The Salyut space stations have been the focus of the Soviet space program since the early 1970s. Like the American astronauts, who use Apollos to ferry them up to Skylab, the Soviets have used Soyuz spacecraft to take them to their Salyut labs. Although both countries have performed similar experiments, the Soviet crew of the Salyut 6 stunned the world with its incredible six months in orbit. If the USSR could do that, said one British scientist, there was

Top: Cosmonaut Alexei Leonov aboard the Soyuz during the link-up with the Apollo. Center: Soyuz spacecraft preparing to dock. Bottom: Apollo spacecraft preparing to dock.

no reason they couldn't also make the trip to Mars.

With space crews from both countries staying in orbit longer and longer, scientists recognized a growing need to develop rescue techniques in case either one's crews encountered major problems. But at first this was impossible. Even though Apollo and Soyuz flights were becoming almost routine, the two systems were utterly different. Crews on Soviet Soyuz craft breathed normal air, whereas Americans on Apollos breathed pure oxygen at one third the pressure. If an astronaut entered a cosmonaut's ship, he would have instantly got a painful case of the "bends." But even entering each other's ships would have been impossible. The two different craft hadn't been designed to dock with each other.

The Apollo-Soyuz Project (ASTP) was meant to correct all this. The problems took three years to solve. In addition to cabin atmosphere differences and docking difficulties, there was the language barrier between the two crews. Both crews trained extensively, with the Soviet cosmonauts visiting Houston and the U.S. astronauts going to Star City, the Soviet space center outside Moscow.

The mission got under way on July 15, 1975, with Tom Stafford, Vance Brand, and Deke Slayton on the Apollo and veteran cosmonauts Alexei Leonov and Valeri Lubasov abroad the two-person Soyuz craft. Two days later, mission commanders Leonov and Stafford shook hands across the hatch connecting the two craft.

"We have succeeded!" said Stafford, in broken Russian. To which Leonov replied, "Good show!"

Despite the success of the Apollo-Soyuz mission, the late 1970s was not a productive period for the American space program. As Washington sought ways to trim its soaring federal deficit, NASA's yearly budget fell from $6 billion to barely $3 billion. From 1975 to 1981, no American went into space.

During this time, though, NASA technicians were developing the Space Shuttle. Originally, it was supposed to be ready for launch in 1979, but it ran into such huge delays and cost overruns that its debut was pushed back to 1981.

The first Shuttle, named Columbia, as was Apollo 11's command module, was the first of a planned fleet of four Shuttles. A monument of advanced technology, it looked like nothing that had ever been in orbit before. Like the shuttles that fol-

The space shuttle blasts off

Specialists on the Space Shuttle

The introduction in 1976 of "Mission Specialist" and "Payload Specialist" designations enabled these non-traditional astronauts to make important contributions to the space program. Due to the relatively large size of the space shuttle cabin, a larger crew was possible, each bringing something new to the last frontier.

DR. SALLY RIDE, joined the Space Program after completing her doctoral degree in astrophysics. She underwent the same rigorous training as her male counterparts and spent two years developing the Remote Manipulator Arm, which the astronauts use to load and unload items from the cargo bay. She was promoted to CAPCOM, or Capsule Communicator, where for two missions she was the "Voice of Houston" before being named flight engineer on the second flight of the space shuttle Challenger. As Flight Engineer, Sally was responsible for monitoring the vital processes which keep the shuttle flying. When asked about the flight she replied "The thing I'll remember most about the flight is that it was fun!"

SENATOR JAKE GARN of Utah spent four months training to be a "human guinea pig" in experiments aboard the space shuttle which were designed to teach scientists more about the causes of "spacesickness". As chairman of the Senate Banking Commission, Senator Garn is in charge of budget allocations for the shuttle program, and was invited by NASA to see firsthand how that money is being used. No stranger to the air, Senator Garn, for many years a member of the Utah Air National Guard, had logged more hours flying jets than any other astronaut except Joe Engle. When asked about his role on the flight, he replied that aside from his duties as guinea pig, he "also did a little cooking" while on board.

DR. GUION S. BLUFORD, aerophysicist, became an astronaut after attaining the rank of colonel in the United States Air Force. Colonel Bluford flew 144 missions over Vietnam during 1966 and 1967, and logged over 3,000 hours flying time before being chosen to conduct experiments with the CFES aboard the space shuttle. The CFES, Continuous Flow Electroporesis System, is an experimental mechanism designed to separate biological materials (such as the different particles which make up human blood) by their surface electrical charges as they are passed through an electrical field.

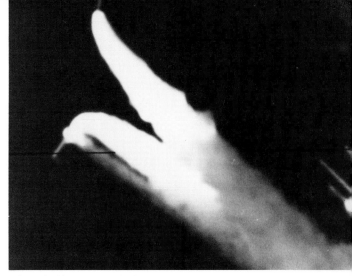

The Challenger explodes on January 28, 1986

lowed it, the Columbia seemed more like a DC-9 airplane. The big difference was that it was attached for take-off to a huge removable fuel tank that was boosted by a pair of removable solid-fuel propellant rockets. Launched as a rocket, it had maneuvered as a spaceship, and then landed as an airplane. With its successful return to earth, scientists predicted that because the new Space Shuttles were reusable, the cost of future forays into space could be cut by as much as 90 percent.

By 1981, space shuttle flights had become almost commonplace. Astronauts conducted experiments, launched new satellites from the shuttle's cargo bay, and even managed to recover two "runaway" satellites at great savings for NASA.

So successful were these missions that NASA announced it would take civilians on upcoming shuttle flights. The first civilian in space would be a teacher so that "all of America will be reminded of the critical role teachers and education play in the life of our nation," declared President Ronald Reagan. In July 1985, Christa McAuliffe, a high school teacher from Concord, New Hampshire, was selected from over 10,000 applicants.

On the morning of January 28, 1986, the Challenger stood on its launch pad at Cape Kennedy, ready for its tenth flight into space. The presence of Christa McAuliffe aboard the shuttle had captured

the nation's attention. As a proud group of family and friends looked on from the nearby grandstands, and millions at home tuned in on their televisions, the Challenger thundered off the launch pad. Within seconds, the ship was ten miles high, moving east at almost 2,000 miles per hour. Suddenly the craft was engulfed in a burst of flame. The Challenger had exploded, killing all seven members of its crew.

As the entire nation mourned, Senator John Glenn, the first American to orbit the earth, put into words what the nation was feeling:

"The explosion of the space shuttle Challenger this morning was nothing less than a tragedy—not only for America's space program but for all Americans. They carried our hopes and dreams with them, let us carry their memory with us."

5

THE GRAND TOUR

Where did our solar system come from? What is it made of? Is there life on some other planet millions of miles from Earth? These are questions that have intrigued scientists and poets for centuries. Yet in spite of the dramatic achievements of the manned space flights during the 1960s, scientists knew even before the first astronauts went into space that the world would be no closer to knowing the answers after those flights than they had been in the past.

There was simply no way that astronauts could find these answers themselves. No spacecraft had been designed that could travel the vast distances from Earth to beyond the moon and on to the planets that lay in the outer reaches of the solar system. The solution would have to

The Voyager 2 spacecraft, traveling through our solar system

be to send unmanned satellites up. And so, even while NASA's manned space flights were being sent to the moon, unmanned satellites—strange-looking but very advanced robots—were probing much farther from Earth.

To achieve success with these robot satellites, NASA had had to tackle some difficult problems. Since each of the nine planets in the solar system orbits the sun at a different speed, launches could take place only when Earth and the particular planet to be explored were in the same relative position. These times are called "launch windows." Also, because the distances to be covered were so enormous, technicians had to create satellites that would last for months and even years.

One of the first plans for exploring the solar system was developed at the Jet Propulsion Laboratory (JPL) in Pasadena, California. In 1963, scientists at JPL came

The planet Venus

up with a plan called the "Grand Tour"—named, in a sense, after the cultured tours of Europe considered an essential part of the education of wealthy American and British young people before the age of air travel. The concept of the Grand Tour was based on the scientific fact that between 1976 and 1980, the five outer planets—Jupiter, Saturn, Uranus, Neptune, and Pluto—would line up in such a way that all five could be explored by a single satellite. Throughout the 1960s, JPL researched and developed the equipment and techniques necessary to carry out their goal. Meanwhile, they also developed much simpler satellites that could fly less ambitious missions. These were the first-generation Mariner probes, designed to fly past a single target within the inner system of planets (Mercury, Venus, and Mars) and send back information.

These early probes were launched by two-stage Atlas booster rockets and had small jet thrusters to stabilize them in space. In most cases, the satellites also had their own engines in case they needed to perform in-flight course corrections. Four winglike panels containing solar satellites powered the satellites' radio and other technical equipment. Since launch windows lasted only a few weeks, the JPL usually sent up two probes during each period, to better the chances of at least one of them being successful.

Between the mid-1960s and the end of 1973, NASA and JPL developed ten separate Mariner probes. Seven of these were successful, sending back to Earth thousands of photographs and new information about our nearest neighbors, Venus, Mars, and Mercury.

Much of the information surprised scientists. Venus, the planet closest to Earth, was discovered to have a surface temperature of 900°F, more than four times hotter than had been expected. With a thick cloud cover hanging over it, it was, as Carl Sagan remarked, "a hell hole of a planet."

Mars was more mysterious. The first successful mission there came within 6,118 miles of the planet and sent back twenty-one photographs. Since these pictures suggested that there was no surface water on Mars, for the first time scientists were able to discount all the science fiction about Martians attacking Earth. After all, without water, how could living beings exist?

The issue of Martian water did not end there, however. In 1969, NASA sent up two satellites at the same time. From just 2,000 feet above Mars's surface, the probes sent back 201 pictures revealing moonlike craters, huge, flat desert areas, mountains, and valleys where no craters existed at all. This fascinated scientists, so they sent up Mariner 9 to probe some more. When this satellite reached Mars, it was greeted by a huge dust storm rising from the surface. Fortunately, this probe had a very sophisticated computer system that allowed Ground Control to reprogram it in flight. The Mariner 9 was ordered not to take pictures until the storm had cleared and the dust had settled. The storm raged on for months. When it finally ended, it was discovered that the heavy amount of dust in the air had caused the surface temperature of the planet to drop more than 36°F. Had the same thing happened on Earth, it might have signaled another ice age. But with the storm now over, Mariner 9 began photographing the planet, taking a total of 7,329 pictures from a distance of 900 miles. These pictures clearly showed the violent history of Mars. There were many volcanoes, one of them three times the size of Mauna Loa, the largest volcano on Earth. Also, on the planet's surface was a canyon four miles deep, stretching the same distance as that from California to Washington, D.C. Most surprising of all, though, were the smaller channels that looked like riverbeds that had been carved by water. Moreover, traces of water were found in Mars's atmosphere. Had the earlier probe been wrong? Was there life on Mars after all?

Last of the Mariner probes was a voyage to Mercury on November 3, 1973. Before reaching that planet, Mariner 10 cruised by Venus, taking the first pictures of it. Then Venus's gravity shot the satellite toward the sun. As a protective shade opened, the craft passed by Mercury—93,000,000 miles away from Earth. Photographs showed that Mercury, like the moon, was heavily cratered and that it had a very thin atmosphere of helium.

In 1975 began an exciting new space-probe project, Viking. Its mission was an attempted landing on Mars. The Viking project consisted of two parts—a Mariner-like craft that would orbit the planet and an instrument capsule that would separate from the orbiter and land on the Martian surface. On August 20, 1975, Viking 1 was launched, followed in September by Viking 2. Ten months later, the first probe entered Mars orbit to look for a landing site. A month after that, the lander separated from the orbiter and began its descent toward a section of Mars called Chryse Planitia, or the Plains of God. Two weeks later, Viking 2 landed on another part of Mars, the Plains of Utopia. Both landings occurred about seven years after American astronauts first stepped out onto the moon.

The first tests performed by the two Vikings were temperature readings. The

Voyager 2 on Mars (note frost in the background)

recorded high was a chilling −24°F—too cold for most forms of life. Mars's atmosphere did contain 3 percent nitrogen, however, which meant that it could have been similar to Earth's atmosphere at an earlier time. Another discovery pointing to the possibility of life was that ice made of water was found underneath the dry ice of the planet's polar caps. Finally, after eight days, Viking 1's lander reached out a mechanical arm and scooped up a mound of Martian soil. Suddenly, scientists felt excited. The soil had water in it! Upon further soil analysis, it was discovered that even though the water was there, no life

forms were present. Disappointing as this evidence was, the Viking probe vastly increased people's knowledge of Mars, sending more than 50,000 pictures back to Earth. What's more, the Viking 1 orbiter continued operating for four years, while the Viking 1 lander continued sending messages from Mars until late 1982.

This first round of Viking probes of the inner planets gave way in 1978 to the Pioneer-Venus missions. The first of these new satellites was launched on May 20, 1978, and reached Venus on December 5. With its mission being to break through the planet's thick cloud cover, Pioneer-Venus used radar beams to measure the planet's surface elevations. The probe discovered that Venus was covered with continent-sized plateaus and huge canyons as well as dormant volcanoes and craters. A second probe, Pioneer-Venus 2, confirmed what had been discovered earlier: The thick atmosphere of Venus could never support life. If some hapless space traveler tried to visit there, he or she would be roasted, suffocated, squashed, and corroded by the intense combination of searing temperatures, atmospheric pressure, and continuous rain of sulphuric acid.

Seven years before Pioneer-Venus 2 had gathered data about Venus, a different set of Pioneers had been launched. Their mission had been to chart a path through the dangerous asteroid belt between Mars and Jupiter and to begin exploring the outer planets. On March 2, 1972, the first of these probes, Pioneer 10 had passed within 82,000 miles of Jupiter. Its partner,

Saturn's rings, which are made up of many smaller rings

Pioneer 11 not only repeated that feat but, on September 1, 1979, came within 13,300 miles of the giant, ringed planet, Saturn.

These two pioneer flights were mere warmups for what was to come next—the Voyager flights had grown out of the Grand Tour idea. Voyagers 1 and 2 were launched in September and August of 1977, with Voyager 2 actually leaving 16 days before its companion. They both reached Jupiter two years later, in 1979, and sent back more than 68,000 pictures, including one that showed a volcano exploding on Jupiter's largest moon, Io. From Jupiter, the two voyagers moved on toward Saturn, arriving two years later. It soon became clear that Saturn was a cold, gaseous ball, chilled by wind blowing at speeds of up to 1,100 mph. Saturn's famous rings proved to be made up of thousands of smaller rings consisting of icy debris. Most important of all, the Voyagers discovered six new moons, bringing Saturn's total to twenty-three. With the completion of the mission, the two Voyagers separated. While Voyager 1 left the solar system, heading for the more distant reaches of the universe, Voyager 2 continued its exploration of our solar system.

In January 1986, Voyager 2 reached the system's seventh planet, Uranus. When

Voyager 2 prepares to leave our solar system

radio signals from Voyager 2 traveled the 1.8 billion miles from Uranus to Earth, scientists learned that the planet was alive with electromagnetic signals and covered with an orange smog. By August of 1989, Voyager 2 is expected to have reached Neptune. Scientists hope it will solve one of Neptune's mysteries—why its moon Triton circles in a direction opposite from the planet's own solar orbit.

Meanwhile, Voyager 1 continues its journey beyond the solar system. Even more extraordinary, though, is the fact that Pioneer 10, launched in 1971, is still traveling through space. On June 13, 1983, twelve years after it left its home planet, it finally left Earth's solar system behind as well and began making its way through the universe.

What will Pioneer 10 find out there? No one knows, of course. In the event that it meets a real extraterrestrial being, it will give whomever it meets a warm greeting from Earth. On its side, it carries a plaque engraved with symbols showing Earth's location in the universe. In addition, there is a picture of a man and a woman. The man is waving hello.

AFTERWORD

"THE THRILL OF JUST BEGINNING"

In the wake of the Challenger tragedy, America's space program seeemd at a standstill. And when a special commission charged by President Reagan to investigate the incident criticized NASA for putting scheduling over safety, it seemed that the space agency would be grounded for some time.

On the morning of September 29, 1988, all that changed. That morning, after almost three years of uncertainty, the space shuttle Discovery soared into orbit. America was back in space.

This time NASA had taken additional safety precautions. First of all, the shuttle had been completely redesigned. Secondly, the Discovery crew, containing only experienced astronauts and space

The space shuttle blasts off on another mission

scientists, trained longer than any crew before them.

The four-day flight, during which the crew launched a new communications satellite from the Discovery's cargo bay, was a complete success. Within just two months, the shuttle Atlantis successfully completed a military mission. In February 1989, a navigational satellite was launched by an unmanned rocket. NASA has announced plans for thirty more shuttle flights before 1992.

As America returned to space, some people again questioned the importance of space exploration. What business did we have spending billions of dollars exploring space, critics asked, when right here on Earth millions of people were living in miserable poverty? Even if the abysmal condition of a large segment of the Earth's population could be ignored, there

were other questions to be considered. With the possibilities of space exploration practically limitless, what space programs should be given priority? And how much money should be spent?

In the beginning of the 1980s, NASA got a big financial boost from Washington. President Ronald Reagan not only increased its funding but also suggested ways that NASA could spend its new money. Among his suggestions was the possible return of astronauts to the moon and perhaps even an attempt to land human beings on Mars. He also proposed opening outer space to business, envisioning such phenomena as orbiting factories. And he announced that he would like to see a permanent space station in orbit by the early 1990s. Not since John F. Kennedy' promise to put men on the moon had any American president made such a sweeping commitment to space research.

Not all of President Reagan's proposals were greeted with enthusiasm. People questioned the cost of such projects, adding that unmanned space flights not only were cheaper but often brought equally worthwhile scientific results.

Whatever the outcome of such debates, there can be no doubt about the benefits that space exploration has given the world so far. It has vastly increased human understanding of the universe and of our own planet. Orbiting satellites have brought people the world over closer together, allowing them to communicate in a matter of seconds. Doctors in New York can get immediate help from doctors in Tokyo when facing emergency, life-threatening situations. Scientists, nutritionists, bankers, and businesspeople across the globe can now work closely together, almost as if they were in the same office. In the future, the almost weightless, germ-free environment of space may lend itself to manufacturing certain drugs. As James Rose, a research director at McDonnell-Douglas, the huge aerospace company, said, "For every new chemical created on Earth, we could make five in space."

More than half a century ago, in 1932, the American rocket pioneer Robert Goddard wrote a letter to H.G. Wells, the essayist and scence-ficton writer. What he said about the continued exploration of space is as applicable today, as we approach the twenty-first century, as it was then.

The space shuttle is escorted back to earth by T-38 chase planes

INDEX

Page numbers in *italics* indicate illustrations

SUGGESTED READING

BERGER, MELVIN. *Space Shots, Shuttles and Satellites.* New York: G.P. Putnam's Sons, 1983.

CROMIE, WILLIAM J. *Skylab.* New York: David McKay Company, Inc., 1976.

FAGET, MAX. *Manned Space Flight.* New York: Holt, Rinehart and Winston, Inc., 1965.

GARDNER, ROBERT. *Space, Frontier of the Future.* Garden City: Doubleday and Company, Inc., 1980.

GURNEY, GENE. *Americans to the Moon, the Story of Project Apollo.* New York: Random House, 1970.

PELLEGRINO, CHARLES R., and JOSHUA STOFF. *Chariots for Apollo.* New York: Atheneum, 1985.

SPARKS, MAJOR JAMES C., USAF (RETIRED). *Moon Landing, Project Apollo.* New York: Dodd, Mead and Co., 1969.

STOCKTON, WILLIAM, and JOHN NOBLE WILFORD. *Spaceliner.* New York: Times Books, 1981.